United States Government Accountability Office

Report to the Committee on Armed Services, U.S. Senate

May 2013

SPECTRUM MANAGEMENT

Federal Relocation Costs and Auction Revenues

GAO-13-472

GAO Highlights

Highlights of GAO-13-472, a report to the Committee on Armed Services, U.S. Senate

May 2013

SPECTRUM MANAGEMENT
Federal Relocation Costs and Auction Revenues

Why GAO Did This Study

Allocating radio-frequency spectrum is a challenging task because of competing commercial and government demands. In 2006, FCC auctioned spectrum licenses in the 1710-1755 MHz band that had previously been allocated for federal use. To meet the continued demand for commercial wireless services, NTIA assessed the viability of reallocating the 1755-1850 MHz band to commercial use; this band is currently assigned to more than 20 federal users, including DOD. In March 2012, NTIA reported that it would cost $18 billion over 10 years to relocate most federal operations from the band, raising questions about whether relocating federal users is a sustainable approach.

GAO was directed to review the costs to relocate federal spectrum users and revenues from spectrum auctions. This report addresses (1) estimated and actual relocation costs, and revenue from the previously auctioned 1710-1755 MHz band; (2) the extent to which DOD followed best practices to prepare its preliminary cost estimate for vacating the 1755-1850 MHz band; and (3) existing government or industry forecasts for revenue from an auction of the 1755-1850 MHz band. GAO reviewed relevant reports; interviewed DOD, FCC, NTIA, and OMB officials and industry stakeholders; and analyzed the extent to which DOD's preliminary cost estimate met best practices as identified in GAO's *Cost Estimating and Assessment Guide (Cost Guide)*. FCC agreed with the report's findings and DOD, FCC, and NTIA provided technical comments that were incorporated as appropriate.

View GAO-13-472. For more information, contact Mark L. Goldstein at (202) 512-2834 or goldsteinm@gao.gov.

What GAO Found

Some federal agencies underestimated the costs to relocate communication systems from the 1710-1755 megahertz (MHz) band, although auction revenues appear to exceed relocation costs by over $5 billion. As of March 2013, actual relocation costs have exceeded estimated costs by about $474 million, or 47 percent. The National Telecommunications and Information Administration (NTIA) expects agencies to complete the relocation effort between 2013 and 2017, with a final relocation cost of about $1.5 billion. Actual relocation costs have exceeded estimated costs for various reasons, including unforeseen challenges and some agencies not following NTIA's guidance for developing cost estimates. However, the Department of Defense (DOD) expects to complete its relocation for about $71 million less than its estimate of about $355 million. NTIA and the Office of Management and Budget (OMB) are taking steps to ensure that agencies improve their cost estimates by, for example, preparing a cost estimation template and guidelines for reporting reimbursable costs. The auction of spectrum licenses in the 1710-1755 MHz band raised almost $6.9 billion.

DOD's preliminary cost estimate for relocating systems out of the 1755-1850 MHz band substantially or partially met GAO's best practices for cost estimates, but changes in key assumptions may affect future costs. Adherence with GAO's *Cost Guide* reduces the risk of cost overruns and missed deadlines. GAO found that DOD's preliminary estimate of $12.6 billion substantially met the comprehensive and well-documented best practices. For instance, it included complete information about systems' life cycles, and the baseline data were consistent with the estimate. However, GAO found that some information on the tasks required to relocate some systems was incomplete. GAO also determined that DOD's estimate partially met the accurate and credible best practices. For example, DOD applied appropriate inflation rates and made no apparent calculation errors. However, DOD did not complete some sensitivity analyses and risk assessments at the program level, and not at all at the summary level. DOD officials said that changes to key assumptions could substantially change relocation costs. Most importantly, decisions about which spectrum band DOD would relocate to are still unresolved, and relocation costs vary depending on the proximity to the 1755-1850 MHz band. Nevertheless, DOD's preliminary cost estimate was consistent with its purpose—informing the decision-making process to make additional spectrum available for commercial wireless services.

No government revenue forecast has been prepared for a potential auction of the 1755-1850 MHz band, and a variety of factors could influence auction revenues. One private sector study in 2011 forecasted $19.4 billion in auction revenue for the band, assuming that federal users would be cleared and the nationwide spectrum price from a previous auction, adjusted for inflation, would apply to this spectrum. Like for all goods, the price of spectrum, and ultimately the auction revenue, is determined by supply and demand. The Federal Communications Commission (FCC) and NTIA jointly influence the amount of spectrum allocated to federal and nonfederal users (the supply). The potential profitability of a spectrum license influences its demand. Several factors would influence profitability and demand, including whether the spectrum is cleared of federal users or must be shared.

United States Government Accountability Office

Contents

Letter		1
	Background	5
	Some Agencies Underestimated 1710-1755 MHz Band Relocation Costs, Although Auction Revenues Appear to Exceed Those Costs	11
	DOD's Preliminary Cost Estimate Substantially or Partially Met GAO's Identified Best Practices, but Changes in Assumptions May Affect Future Costs	17
	No Government Revenue Forecasts Exist for a Potential Auction of the 1755-1850 MHz Band, and a Variety of Factors Could Influence Auction Revenues	23
	Agency Comments	27
Appendix I	Objectives, Scope, and Methodology	29
Appendix II	Comments from the Federal Communications Commission	32
Appendix III	GAO Contact and Staff Acknowledgments	33
Tables		
	Table 1: Comparison of Estimated and Actual Relocation Costs for the 1710-1755 MHz Band (as of March 2013)	12
	Table 2: Summary Assessment of DOD Spectrum Relocation Cost Estimate Compared to GAO-Identified Best Practices	19
Figure		
	Figure 1: Examples of Allocated Spectrum Uses and DOD Systems Using the 1755-1850 MHz Band	6

Abbreviations

AWS	Advanced Wireless Services
AWS-1	Advanced Wireless Services-1 auction
CAPE	Cost Assessment and Program Evaluation
CBO	Congressional Budget Office
Cost Guide	*Cost Estimating and Assessment Guide*
CSEA	Commercial Spectrum Enhancement Act
DOD	Department of Defense
FCC	Federal Communications Commission
GHz	gigahertz
ICE	Immigration and Customs Enforcement
kHz	kilohertz
MHz	megahertz
MHz-pop	megahertz-population
NTIA	National Telecommunications and Information Administration
OMB	Office of Management and Budget
Wi-Fi	wireless fidelity

This is a work of the U.S. government and is not subject to copyright protection in the United States. The published product may be reproduced and distributed in its entirety without further permission from GAO. However, because this work may contain copyrighted images or other material, permission from the copyright holder may be necessary if you wish to reproduce this material separately.

GAO

U.S. GOVERNMENT ACCOUNTABILITY OFFICE

441 G St. N.W.
Washington, DC 20548

May 22, 2013

The Honorable Carl Levin
Chairman
The Honorable James M. Inhofe
Ranking Member
Committee on Armed Services
United States Senate

As the demand for and use of smart phones, tablets, and other wireless devices continues to grow and new mission needs unfold among federal government agencies, nearly all parties are becoming increasingly concerned about the availability of radio frequency spectrum to meet future commercial and federal needs.[1] Federal users require spectrum for national defense, homeland security, and other vital mission activities. While the national interest may be served by a robust commercial wireless broadband system, the federal government also needs spectrum to support critical missions, including military operations, testing, and training at home and around the world. Thus, balancing competing industry and government demands for a limited amount of spectrum, today and in the future, is a challenging and complex task.

To address these challenges, current and past administrations and Congresses, the Federal Communications Commission (FCC), and other stakeholders have proposed various policy, economic, and technological solutions to address the availability and efficient use of spectrum. For example, in 2010, the Obama administration issued a presidential memorandum directing the National Telecommunications and Information Administration (NTIA) within the Department of Commerce to collaborate with FCC to develop a plan and a timetable to make 500 megahertz (MHz) of federally and nonfederally allocated spectrum available for

---

[1]The radio frequency spectrum is the part of the natural spectrum of electromagnetic radiation lying between the frequency limits of 3 kilohertz (kHz) and 300 gigahertz (GHz). Radio frequencies are grouped into bands and are measured in units of Hertz, or cycles per second. The term kHz refers to thousands of Hertz, megahertz (MHz) to millions of Hertz, and GHz to billions of Hertz. The Hertz unit of measurement is used to refer to both the quantity of spectrum (such as 500 MHz of spectrum) and the frequency bands (such as the 1755-1850 MHz band).

wireless broadband use in the next 10 years.[2] The majority of this freed-up spectrum would likely be auctioned and licensed for mobile broadband and other high-value commercial uses.[3] As part of this effort, NTIA led an 8-month interagency evaluation process to determine whether it would be possible to repurpose 95 MHz in the 1755-1850 MHz band for commercial wireless services. The 1755-1850 MHz band is ideally suited for both commercial and federal users because its radio wave characteristics enable highly mobile, yet reliable communication links.[4] Within the United States, this band is currently allocated exclusively to the federal government, particularly for defense purposes, such as military tactical communications, air combat training, and space systems.

In March 2012, NTIA reported that the preliminary cost estimate to relocate most federal operations from the 1755-1850 MHz band would be about $18 billion over 10 years.[5] Approximately $12.6 billion of this estimate was attributed to relocating military systems. The report was largely based on inputs from the federal agencies using the band, including the Department of Defense (DOD). NTIA's analysis showed that while there are significant challenges to overcome, it might be possible to free up all 95 MHz of spectrum in the 1755-1850 MHz band. The report describes the required conditions to ensure no loss of critical capabilities for the more than 20 federal agencies operating a variety of systems in

---

[2]Memorandum for the Heads of Executive Departments and Agencies, Unleashing the Wireless Broadband Revolution (Presidential Memorandum), 75 Fed Reg. 38387 (June 28, 2010).

[3]To promote more efficient use of spectrum and meet future needs, FCC has increasingly adopted more market-oriented approaches to spectrum management in recent years, including using a competitive bidding process, or auctions, to assign spectrum licenses to commercial users. Since 1997, with certain limited exceptions, Congress has *required* FCC to assign licenses by auction in situations where it permits the filing of mutually exclusive applications. 47 U.S.C. § 309(j). From 1994, when FCC first implemented its auction authority, through January 2013, FCC held 81 auctions and generated nearly $52 billion for the U.S. Treasury.

[4]The wavelength of a frequency is a key determinant of its best uses. Frequencies above about 3 GHz are not as conducive to mobile communications as are lower frequencies that require less energy to transmit signals over a given distance and are more capable of penetrating walls and buildings. See Coleman Bazelon, The Brattle Group, *The Economic Basis of Spectrum Value: Pairing AWS-3 with the 1755 MHz Band is More Valuable than Pairing it with Frequencies from the 1690 MHz Band* (Washington, D.C.: Apr. 11, 2011).

[5]See NTIA, *An Assessment of the Viability of Accommodating Wireless Broadband in the 1755-1850 MHz Band* (Washington, D.C.: March 2012).

that band at any given time across the nation.[6] Relocating to other parts of the radio frequency spectrum means that many of these existing systems would need to be redesigned. The high cost and lengthy time to implement this relocation has raised questions about whether an auction of spectrum licenses in the 1755-1850 MHz band would cover those expenses, as required by federal law,[7] and whether other approaches, such as sharing spectrum between commercial and federal users, would better achieve spectrum policy goals.

Because of the importance of this issue to private and public interests in the United States, the Senate Armed Services Committee directed us to review the historical differences between estimated federal relocation costs and actual auction revenue, and assess whether the cost of vacating or sharing subsets of the 1755-1850 MHz band is sufficiently captured in preliminary cost estimates.[8] To address these issues, we examined (1) the differences, if any, between estimated and actual federal relocation costs and revenue from the auction of the 1710-1755 MHz band, (2) the extent to which DOD followed best practices to prepare its preliminary cost estimate for vacating the 1755-1850 MHz band and the limitations, if any, of its analysis, and (3) what government or industry revenue forecasts exist for an auction of the 1755-1850 MHz band, and what factors, if any, could influence the actual auction revenue. We provided a preliminary briefing to your offices on February 25, 2013; this report formally transmits our findings.

To examine the differences between federal relocation costs and revenue from the auction of the 1710-1755 MHz band, we reviewed recent NTIA

---

[6]The federal agencies with operations in the 1755-1850 MHz band include, among others, the Departments of Commerce, Defense, Energy, Health and Human Services, Homeland Security, Housing and Urban Development, Interior, Justice, Treasury, and Veterans Affairs, and the Federal Aviation Administration, National Aeronautics and Space Administration, Office of Personnel Management, U.S. Agency of International Development, U.S. Capitol Police, and U.S. Postal Service. The Department of Defense includes the Air Force, Army, Navy, and Marine Corps.

[7]The Communications Act, Act of June 19, 1934, ch. 652, Title I, §§ 8, 9 (the Communications Act), as amended by the Commercial Spectrum Enhancement Act (CSEA), Pub. L. No. 108-494, § 202, 118 Stat. 3986, (2004), codified at 47 U.S.C. § 309(j)(3)(F), requires that auction proceeds must be equal to at least 110 percent of the total estimated relocation costs for FCC (the agency conducting the auction) to conclude the auction. 47 U.S.C. § 309(j)(16)(B).

[8]S. Rept. 112-173 (2012).

spectrum relocation reports, including NTIA's annual progress reports for the 1710-1755 MHz transition; FCC spectrum auction data as of December 2012; and the Office of Management and Budget's (OMB) 2007 report to Congress on Agency Plans for Spectrum Relocation Funds. To assess the reliability of the relocation cost and auction revenue data, we reviewed documentation related to the data, compared the data to other sources, including government reports, and discussed the data with FCC and NTIA officials. We did not evaluate the accuracy of individual agencies' relocation cost data, as this was outside the scope of our review. Based on this review, we determined that the FCC and NTIA data were sufficiently reliable for the purposes of our report. To determine the number of auctions involving the relocation of federal agencies to new spectrum frequencies, we reviewed FCC auction data and NTIA reports, and interviewed officials from FCC, NTIA, OMB, and the Congressional Budget Office (CBO); the Advanced Wireless Services-1 (AWS-1) auction involving the 1710-1755 MHz band was the only spectrum auction involving federal agencies with significant, known relocation costs.[9]

To assess whether the cost of vacating the 1755-1850 MHz band is sufficiently captured in DOD's preliminary cost estimate, we compared DOD's preliminary estimate against the best practices in GAO's *Cost Estimating and Assessment Guide* (*Cost Guide*).[10] The *Cost Guide* identifies best practices that help ensure cost estimates are comprehensive, well-documented, accurate, and credible, and it has been used to evaluate cost estimates across the government. To identify any limitations affecting DOD's estimate, we also interviewed DOD officials responsible for developing the department's preliminary cost estimate. To identify any government or industry forecasts of revenue from a future auction of the 1755-1850 MHz band and any factors that would affect the value of spectrum licenses, we reviewed academic, government, and public policy literature, focusing on studies mentioning (1) spectrum

---

[9]There have been other auctions involving the relocation of federal government agencies. For example, the National Oceanic and Atmospheric Administration, the Air Force, and the National Science Foundation previously operated systems in the 1670-1675 MHz band. The estimated cost to relocate these systems was $35-55 million for the National Oceanic and Atmospheric Administration and $515,000 for the Air Force. See NTIA, *Spectrum Reallocation Final Report: Response to Title VI – Omnibus Budget Reconciliation Act of 1993* (Washington, D.C.: February 1995). FCC auctioned the band in April 2003, and the auction generated $12.6 million. Final relocation costs are unclear.

[10]GAO, *GAO Cost Estimating and Assessment Guide: Best Practices for Developing and Managing Capital Program Costs*, GAO-09-3SP (Washington, D.C.: March 2009).

auctions involving the relocation of federal agencies, (2) spectrum valuation and revenues from the sale of spectrum licenses, and (3) relocation costs. We also interviewed officials from CBO and OMB, and stakeholders with knowledge of spectrum licensing issues, including industry and policy experts. For more details on our scope and methodology, see appendix I.

We conducted this performance audit from September 2012 to May 2013 in accordance with generally accepted government auditing standards. Those standards require that we plan and perform the audit to obtain sufficient, appropriate evidence to provide a reasonable basis for our findings and conclusions based on our audit objectives. We believe that the evidence obtained provides a reasonable basis for our findings and conclusions based on our audit objectives.

Background

The radio frequency spectrum is the resource that makes possible wireless communication and supports a vast array of commercial and government services. Federal, state, and local agencies use spectrum to fulfill a variety of government missions, such as national defense, air-traffic control, weather forecasting, and public safety. DOD uses spectrum to transmit and receive critical voice and data communications involving military tactical radio, air combat training, precision-guided munitions, unmanned aerial systems, and aeronautical telemetry and satellite control, among others. The military employs these systems for training, testing, and combat operations throughout the world. Commercial entities use spectrum to provide a variety of wireless services, including mobile voice and data, paging, broadcast television and radio, and satellite services.

In the United States, responsibility for spectrum management is divided between two agencies: FCC and NTIA. FCC manages spectrum for nonfederal users, including commercial, private, and state and local government users, under the Communications Act.[11] NTIA manages spectrum for federal government users and acts for the President with respect to spectrum management issues as governed by the National Telecommunications and Information Administration Organization Act.[12]

---

[11] 47 U.S.C. § 309.

[12] Pub. L. No. 102-538, title I, 106 Stat. 3533, codified as amended at 47 U.S.C. ch. 8.

FCC and NTIA manage the spectrum through a system of frequency allocation and assignment.

- *Allocation* involves segmenting the radio spectrum into bands of frequencies that are designated for use by particular types of radio services or classes of users. (Fig. 1 illustrates examples of allocated spectrum uses, including DOD systems using the 1755-1850 MHz band.) In addition, spectrum managers specify service rules, which include the technical and operating characteristics of equipment.

Figure 1: Examples of Allocated Spectrum Uses and DOD Systems Using the 1755-1850 MHz Band

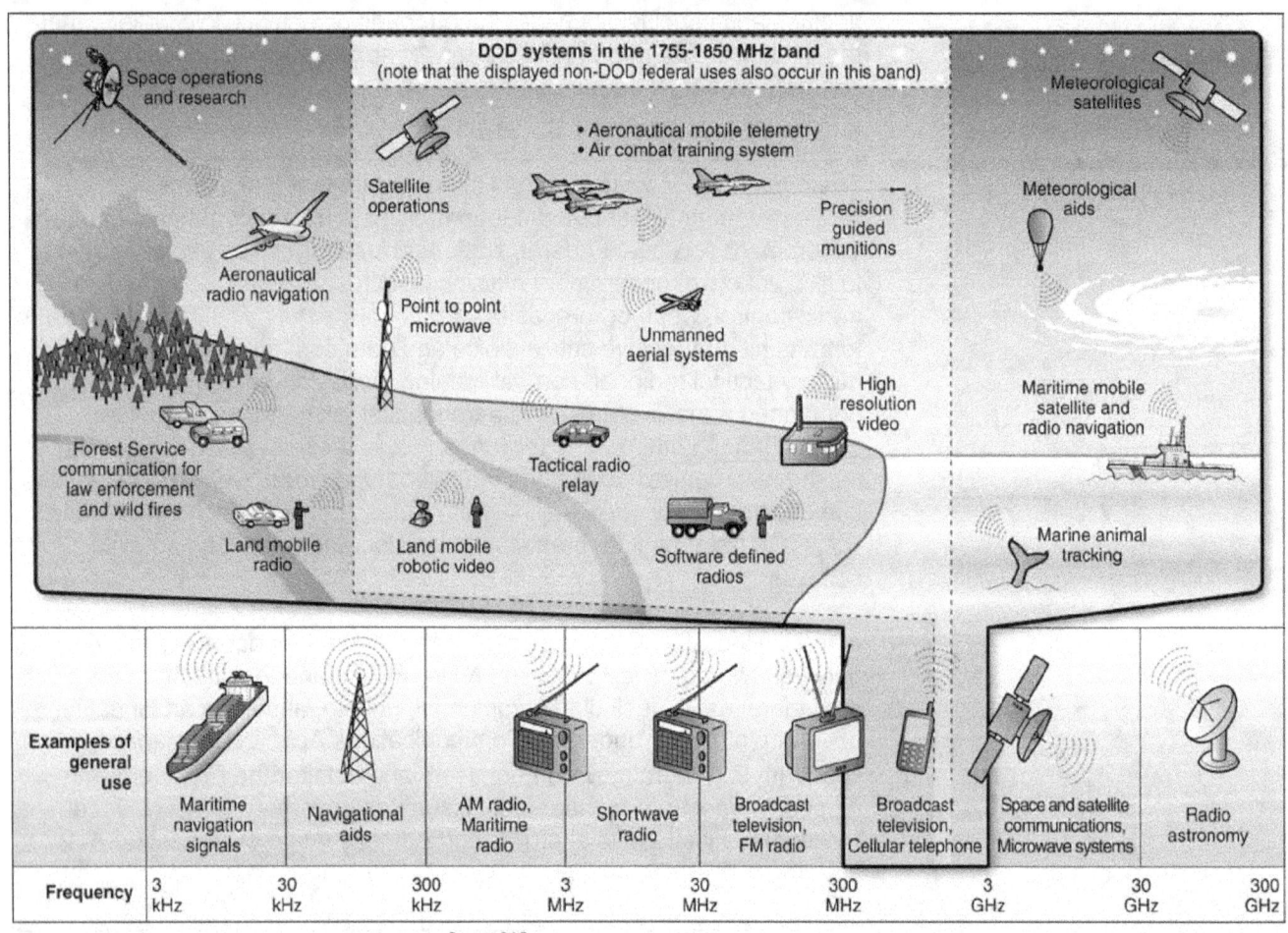

Source: GAO.

- *Assignment*, which occurs after spectrum has been allocated for particular types of services or classes of users, involves providing users, such as commercial entities or government agencies, with a license or authorization to use a specific portion of spectrum. FCC assigns licenses within frequency bands to commercial enterprises, state and local governments, and other entities. Since 1994, FCC has used competitive bidding, or auctions, to assign certain licenses to commercial entities for their use of spectrum.[13] Auctions are a market-based mechanism in which FCC assigns a license to the entity that submits the highest bids for specific bands of spectrum. NTIA authorizes spectrum use through frequency assignments to federal agencies. More than 60 federal agencies and departments combined have over 240,000 frequency assignments across all spectrum bands, although 9 departments, including DOD, hold 94 percent of all frequency assignments for federal use.

Congress has taken a number of steps to facilitate the deployment of innovative, new commercial wireless services to consumers, including requiring more federal spectrum to be reallocated for commercial use. Previously, reallocation of spectrum from federal to private-sector users was directed under the Omnibus Budget Reconciliation Act of 1993, which was later expanded by the Balanced Budget Act of 1997.[14] Relocating federal communications systems to other spectrum bands to accommodate private sector activities can involve significant capital investment costs. The cost of relocating communications systems is affected by many variables related to the systems themselves as well as the relocation plans. Some fixed microwave systems, for example, can generally use off-the-shelf commercial technology or may just need to be re-tuned to accommodate a change in frequency. However, some systems may require significant modification if the characteristics of the new spectrum frequencies differ sufficiently from the original spectrum. Specialized systems, such as those used for surveillance and law

---

[13] As noted above, not all licenses are assigned via auctions. For example, auctions are precluded for public safety and noncommercial broadcast stations. 47 U.S.C. § 309(j)(2). FCC allocates some frequency bands for unlicensed use—that is, users do not need to obtain a license to use the spectrum. Rather, an unlimited number of unlicensed users share those frequencies. Thus, the assignment process does not apply to unlicensed spectrum. Similarly, auctions are not used where FCC relies on licensing processes that do not contemplate the filing of competing applications (e.g., first-come, first-served procedures or situations where commercial users can share spectrum).

[14] Pub. L. No. 103-66, § 6001, 107 Stat. 312 (1993) (OBRA-93) amended by Pub. L. No. 105-33, § 3002, 111 Stat. 251(1997) (BBA-97), codified as amended at 47 U.S.C. § 923.

enforcement purposes, may not be compatible with commercial technology, and therefore agencies have to work with vendors to develop equipment that meets mission needs and operational requirements.

In 2004, the Commercial Spectrum Enhancement Act (CSEA) established a Spectrum Relocation Fund,[15] funded from auction proceeds, to cover the costs incurred by federal entities that relocate to new frequency assignments or transition to alternative technologies.[16] OMB administers the Spectrum Relocation Fund in consultation with NTIA. CSEA streamlined the process by which federal agencies are reimbursed for relocation costs and requires FCC to notify NTIA at least 18 months in advance of beginning an auction of new licenses of spectrum identified for reallocation from federal to nonfederal use. It also requires NTIA to provide estimated cost and transition timing data to FCC, Congress, and GAO at least 6 months prior to the auction, and requires that auctions recover at least 110 percent of these estimated costs. CSEA was amended by the Middle Class Tax Relief and Job Creation Act of 2012, further easing relocation by (1) allowing agencies to use some of the funding for advance planning and system upgrades, (2) extending the reimbursement scheme to sharing as well as relocation expenses, and (3) requiring agencies to submit transition plans for relocation (or sharing) for interagency management review of the costs and timelines associated with the relocation.[17]

The auction of spectrum licenses in the 1710-1755 MHz band was the first with relocation costs to take place under CSEA. CSEA designated 1710-1755 MHz as "eligible frequencies" for which federal relocation costs could be paid from the Spectrum Relocation Fund, which is funded by the proceeds from the auction of the band.[18] Twelve federal agencies previously operated communications systems in this band, including

---

[15] 47 U.S.C. § 928.

[16] Eligible relocation expenses are those costs incurred by a federal entity to achieve comparable capability of systems, regardless of whether that is achieved by relocating to a new frequency assignment or using an alternative technology. 47 U.S.C. § 923(g)(3).

[17] Pub. L. No. 112-96, § 6701(a)(1)(D), 126 Stat. 156, 246-247 (2012), codified at 47 U.S.C. § 923(g)(3)(A), (B)(ii).

[18] 47 U.S.C. § 923(g)(2).

DOD.[19] NTIA and FCC jointly reallocated the 1710-1755 MHz band for nonfederal use, and FCC designated the spectrum for Advanced Wireless Services (AWS).[20] In September 2006, FCC concluded the AWS-1 auction of licenses in the 1710-1755 MHz band.[21] In accordance with CSEA,[22] a portion of the auction proceeds associated with the 1710-1755 MHz band is currently being used to pay spectrum relocation expenses.

In addition to the 1710-1755 MHz band, the wireless industry has expressed interest in the 1755-1850 MHz band, largely because the band offers excellent radio wave propagation, enabling mobile communication links.[23] The federal government has studied the feasibility of relocating federal agencies from the 1755-1850 MHz band on several occasions. For example, in March 2001, NTIA issued a report examining the potential to accommodate mobile wireless services in the broader 1710-1850 MHz band. The report was largely based on input from other federal agencies, including a DOD study. NTIA found that unrestricted sharing of the 1755-1850 MHz band was not feasible and that considerable coordination between industry and DOD would be required before any wireless systems could operate alongside federal systems in the band. In August 2001, we also found that more analysis was needed to support spectrum use decisions in the 1755-1850 MHz band, largely because major considerations either were not addressed or were not adequately

---

[19]The 12 federal agencies with communications systems in the 1710-1755 MHz band were the Departments of Agriculture, Defense, Energy, Homeland Security, Housing and Urban Development, Interior, Justice, Transportation, Treasury, the National Aeronautics and Space Administration, the Tennessee Valley Authority, and the U.S. Postal Service.

[20]*In the Matter of Amendment of Part 2 of the Commission's Rules to Allocate Spectrum below 3 GHz for Mobile and Fixed Services*, 17 FCC Rcd. 23193 (2002) states that AWS is the collective term used by FCC "for new and advanced wireless applications, such as voice, data, and broadband services provided over a variety of high-speed fixed and mobile networks."

[21]The AWS-1 auction included licenses in the 1710-1755 MHz and 2110-2155 MHz bands. In August 2008, FCC held a second auction of the AWS-1 licenses that were not sold in the first auction.

[22]47 U.S.C. § 928(d)(1) appropriates from the Spectrum Relocation Fund such sums as may be required to pay authorized relocation or sharing costs. See also 47 U.S.C. § 928(c).

[23]Industry stakeholders have also expressed interest in the 1755-1780 MHz portion of the 1755-1850 MHz band. Some would like to see the 1755-1780 MHz band paired with the 2155-2180 MHz band, which FCC must auction by February 22, 2015.

addressed in DOD's study.[24] These considerations included complete technical and operation analyses of anticipated spectrum interference; cost estimates supporting DOD reimbursement claims; spectrum requirements supporting future military operations; programmatic, budgeting, and schedule decisions needed to guide analyses of alternatives; and potential effects of U.S. reallocation decisions upon international agreements and operations. At the end, a decision was made to reallocate just the 1710-1755 MHz band to minimize the impact on federal capabilities. Activity surrounding the rest of the band (i.e., the 1755-1850 MHz band) did not resurface until October 2010 when NTIA's Fast Track study identified the band for possible reallocation.[25]

In June 2010, the administration issued a presidential memorandum titled "Unleashing the Wireless Broadband Revolution" directing NTIA to collaborate with FCC to make a total of 500 MHz of federal and nonfederal spectrum available for wireless broadband within 10 years. Responding to the President's initiative, in October 2010, NTIA published a plan and timetable to make available 500 MHz of spectrum for wireless broadband. This plan and timetable specified that candidate bands would be prioritized for detailed evaluation to determine the feasibility of vacating the bands to accommodate wireless services. In January 2011, NTIA selected the 1755-1850 MHz band as the priority band for detailed evaluation for relocation. DOD and other affected agencies provided NTIA their input on the spectrum feasibility study for the 1755-1850 MHz band, and NTIA subsequently issued its assessment of the viability for accommodating commercial wireless broadband in the band in March 2012.[26] Most recently, the President's Council of Advisors on Science and Technology published a report in July 2012 recommending specific steps to ensure the successful implementation of the President's 2010

---

[24]GAO, *Defense Spectrum Management: More Analysis Needed to Support Spectrum Use Decisions for the 1755-1850 MHz Band*, GAO-01-795 (Washington, D.C.: Aug. 20, 2001).

[25]See NTIA, *An Assessment of the Near-Term Viability of Accommodating Wireless Broadband Systems in the 1675-1710 MHz, 1755-1780 MHz, 3500-3650 MHz, and 4200-4220 MHz, 4380-4400 MHz Band* (Washington, D.C.: October 2010).

[26]Most of these inputs are listed at http://www.ntia.doc.gov/report/2012/assessment-viability-accommodating-wireless-broadband-1755-1850-mhz-band. DOD's inputs have not been approved for public release.

memorandum.[27] The report found, for example, that clearing and vacating federal users from certain bands was not a sustainable basis for spectrum policy largely because of the high cost to relocate federal agencies and disruption to federal missions. The report recommended new policies to promote the sharing of federal spectrum. The sharing approach has been questioned by CTIA–The Wireless Association and its members,[28] which argue that cleared spectrum and an exclusive-use approach to spectrum management has enabled the U.S. wireless industry to invest hundreds of billions of dollars to deploy mobile broadband networks resulting in economic benefits for consumers and businesses.

Some Agencies Underestimated 1710-1755 MHz Band Relocation Costs, Although Auction Revenues Appear to Exceed Those Costs

Some Federal Agencies Underestimated Relocation Costs

Actual costs to relocate communications systems for 12 federal agencies from the 1710-1755 MHz band have exceeded original estimates by about $474 million, or 47 percent, as of March 2013. Table 1 compares estimated relocation costs with the actual costs based on funds transferred to federal agencies in support of the 1710-1755 MHz band relocation effort. OMB and NTIA officials expect the final relocation cost to be about $1.5 billion compared with the original estimate of about

---

[27] Executive Office of the President, President's Council of Advisors on Science and Technology, *Report to the President: Realizing the Full Potential of Government-Held Spectrum to Spur Economic Growth* (Washington, D.C.: July 2012).

[28] CTIA-The Wireless Association is an international nonprofit membership organization that has represented the wireless communications industry since 1984. Membership in the association includes wireless carriers and their suppliers, as well as providers and manufacturers of wireless data services and products.

$1 billion. In addition, NTIA expects agencies to complete the relocation effort between 2013 and 2017.[29]

Table 1: Comparison of Estimated and Actual Relocation Costs for the 1710-1755 MHz Band (as of March 2013)

Department/agency	Estimated relocation costs[a]	Current actual relocation costs[b]
Agriculture	$21,578,486	$21,578,486
Defense	355,351,524	289,846,448
Energy	176,820,959	212,200,959
Homeland Security	89,994,832	282,239,840
Housing and Urban Development	21,115	21,115
Interior	25,411,949	31,936,326
Justice	262,821,000	556,424,000
Transportation	58,062,020	58,062,020
Treasury	5,301,000	5,301,000
National Aeronautics and Space Administration	740,000	740,000
Tennessee Valley Authority	10,687,857	15,751,057
United States Postal Service	1,761,760	8,333,760
Total	**1,008,552,502**	**1,482,435,011**

Source: NTIA, Relocation of Federal Radio Systems from the 1710-1755 MHz Spectrum Band: Sixth Annual Progress Report (Washington, D.C.: March 2013).

[a]Estimated relocation costs are based on OMB's 2007 report to Congress. See OMB, Commercial Spectrum Enhancement Act: Report to Congress on Agency Plans for Spectrum Relocation Funds (Washington, D.C.: Feb. 16, 2007).

[b]Current actual relocation costs are based on the total amount provided to the agencies from the Spectrum Relocation Fund.

The original transfers from the Spectrum Relocation Fund to agency accounts were made in March 2007. Subsequently, some agencies requested additional monies from the Spectrum Relocation Fund to cover relocation expenses. Agencies requesting the largest amounts of subsequent transfers include the Department of Justice ($294 million), the Department of Homeland Security ($192 million), and the Department of

---

[29]As of December 2012, all eligible systems have ceased operations in the 1710-1755 MHz band. However, some agencies continue to spend Spectrum Relocation Fund monies to implement their new communications systems towards achieving comparable capability.

Energy ($35 million). Total actual costs for the 1710-1755 MHz transition exceeded estimated costs, as reported to Congress in 2007, for many reasons, including:

- **Unforeseen challenges:** Agencies encountered various unforeseen challenges when relocating systems out of the 1710-1755 MHz band. For example, according to NTIA officials, one agency needed to upgrade its radio towers to comply with new standards adopted after the towers were built. The agency requested additional monies from the Spectrum Relocation Fund to cover the cost of upgrading its towers, which had not been part of the agency's original relocation estimate.

- **Unique issues posed by specific equipment location:** According to NTIA, some federal government communications systems are located in remote areas. One agency requested additional monies from the Spectrum Relocation Fund to use a helicopter to replace a fixed microwave system located on a mountain-top, which exceeded its original cost estimate.

- **Administrative issues associated with transition time frame:** NTIA officials told us that some agencies experienced higher than expected labor costs during the transition period, partly to accommodate auction winners' requests to vacate the spectrum as quickly as possible.

- **Costs associated with achieving comparable capability:** Some communications systems are unique to federal agencies, making them difficult to upgrade or relocate. In some instances, agencies were using analog radio systems throughout the 1710-1755 MHz band and the digital technology needed to achieve comparable capability was not available prior to vacating the band. When the technology did become available, some agencies found they needed additional funds to procure it, according to OMB officials. For example, we previously reported that the Department of Justice requested funds exceeding its estimate to develop new technology that would operate using the new spectrum and match its current capabilities.[30]

---

[30]GAO, *Assessment of the Explanation That the Department of Justice Provided for Its Subsequent Transfer from the Spectrum Relocation Fund* (Washington, D.C.: Apr. 23, 2009).

- **Some agencies might not have followed guidance:** Some agencies may not have properly followed OMB and NTIA guidance in preparing their original cost estimates. For instance, Immigration and Customs Enforcement (ICE) did not detail its estimated costs by equipment, location, systems, or frequency as suggested by NTIA's guidance. Instead, the agency provided a lump sum estimate for its spectrum relocation costs. We previously reported that ICE officials did not identify a significant number of relocation expenses in the agency's original transfer request, including costs associated with additional equipment, offices, and systems, among other items.[31] Moreover, according to OMB staff, the agency's initial estimate was based on an inadequate inventory of deployed systems.[32]

Although 11 of the 12 agencies plan to spend the same amount or more than they estimated, DOD expects to complete the 1710-1755 MHz transition for about $284 million, or approximately $71 million less than the original estimated cost of about $355 million.[33] DOD officials told us that the relocation of systems from the 1710-1755 MHz band has been less expensive than originally estimated because it was possible to re-tune many federal systems to operate in the 1755-1850 MHz band and still meet federal mission requirements. DOD's cost estimates, some made as early as 1995, changed over time as officials considered different relocation scenarios with differing key assumptions, and their thinking evolved about the federal systems that would be affected, according to DOD and NTIA officials. Cost estimates to relocate military systems from the late 1990s and early 2000s ranged from a low of $38 million to as much as $1.6 billion, depending on the scenario. For example, the $38 million estimate included costs primarily to relocate and

---

[31]GAO, *Assessment of the Explanation That Immigration and Customs Enforcement Provided for Its Subsequent Transfer from the Spectrum Relocation Fund*, GAO-08-846R (Washington, D.C.: Sept. 8, 2008).

[32]We recommended that OMB, in consultation with NTIA, establish written criteria for agencies to follow when submitting supplemental requests for spectrum relocation funding, communicate these criteria to each of the affected federal entities, and require that these criteria be followed before OMB approves subsequent requests. OMB implemented this recommendation by developing guidance applicable to agencies receiving funds from the Spectrum Relocation Fund. See GAO-08-846R.

[33]To date, the Department of the Navy has initiated the process to return about $65 million to the Spectrum Relocation Fund, as its relocation costs may end up being less than expected. The Department of the Navy is still in the process of finalizing relocation of its systems, and the exact amount of any money that may be returned will not be known until the relocation is complete.

re-tune fixed microwave systems from the 1710-1755 MHz band into the adjacent 1755-1850 MHz band, and it assumed exclusion zones—geographic areas where commercial licensees could not operate—around 16 DOD sites to prevent interference from commercial users. DOD also estimated a cost of an additional $100 million if precision guided munitions operations needed to be relocated from the 1755-1850 MHz band.[34] Subsequently, in 2001, NTIA reported additional cost estimates reflecting several other options under consideration.[35] One option, which was not evaluated by DOD, included a preliminary cost figure of $1.6 billion. This estimate was based on eliminating some of the 16 exclusion zones around DOD sites and, therefore, relocating additional systems that were not included in the original estimate of $38-138 million, according to NTIA. In December 2006, NTIA reported that DOD's estimate to relocate systems would be about $355.4 million. This estimate reflected a new set of assumptions, such as maintaining exclusion zones at 2 of the 16 DOD sites and relocating fixed microwave systems to the 1755-1850 MHz portion of the band or to other federal bands.

Both NTIA and OMB are taking steps to ensure that agencies improve their cost estimates for a future relocation from the 1755-1850 MHz band. For example, according to NTIA and OMB officials, the agencies prepared a cost estimation template and guidelines for reimbursable costs as part of the process to estimate relocation costs for the 1755-1850 MHz band. The Middle Class Tax Relief and Job Creation Act of 2012 expanded the types of costs for which federal agencies can receive payments from the Spectrum Relocation Fund. The act permits agencies to receive funds for costs associated with planning for FCC auctions and studies or analyses conducted in connection with relocation or sharing of spectrum, including coordination with auction winners.[36] In November 2012, OMB issued guidance to federal agencies to clarify allowable pre-auction costs and other requirements that are eligible to receive

---

[34]NTIA, *Assessment of Electromagnetic Spectrum Reallocation: Response to Title X of the National Defense Authorization Act for Fiscal Year 2000*, NTIA-01-44 (Washington, D.C.: January 2001).

[35]NTIA, *The Potential for Accommodating Third Generation Mobile Systems in the 1710-1850 MHz Band: Federal Operations, Relocation Costs, and Operation Impacts*, NTIA-01-46 (Washington, D.C.: March 2001).

[36]Pub. L. No. 112-96, § 6701(a)(1)(D), 126 Stat. 156, 246-247 (2012), codified at 47 U.S.C. § 923(g)(3)(A), (B)(ii).

payments from the Spectrum Relocation Fund.[37] NTIA and OMB officials stated that they are optimistic that by providing pre-auction planning funds to agencies, future cost estimates will improve.

Auction Revenues Appear to Exceed Agency Relocation Costs

The Advanced Wireless Services auction of the 1710-1755 MHz band raised almost $6.9 billion in gross winning bids from the sale of licenses to use these frequencies.[38] Our analysis of auction revenue compared to actual relocation costs suggests that the auction of the 1710-1755 MHz band raised $5.4 billion for the U.S. Treasury. This number reflects the difference between the $6.9 billion auction revenue and the approximately $1.5 billion estimated final federal relocation cost. As mentioned above, NTIA reports that it expects agencies to complete the relocation effort between 2013 and 2017; therefore the final net revenue amount may change. For example, some agencies have returned or plan to return excess relocation funds to the Spectrum Relocation Fund.

---

[37]OMB, *Guidance for Agencies on Transfers from the Spectrum Relocation Fund for Certain Pre-Auction Costs*, M-13-01 (Washington, D.C.: Nov. 20, 2012).

[38]Although the AWS-1 auction of spectrum licenses raised $13.7 billion, the portion of the auction proceeds associated with the transferred government spectrum amounted to almost $6.9 billion and was deposited in the Spectrum Relocation Fund.

DOD's Preliminary Cost Estimate Substantially or Partially Met GAO's Identified Best Practices, but Changes in Assumptions May Affect Future Costs

DOD's Preliminary Cost Estimate for Relocating from the 1755-1850 MHz Band Substantially or Partially Met GAO's Identified Best Practices

To prepare the preliminary cost estimate portion of its study to determine the feasibility of relocating DOD's 11 major radio systems from the 1755-1850 MHz band, DOD officials said the agency implemented the following methodology:

- DOD's Cost Assessment and Program Evaluation (CAPE) group[39] led the effort and provided guidance to management at the respective military services regarding the data needed to support each system's relocation cost estimate and how they should be gathered to maintain consistency across the services. The guidance used by CAPE was based on guidance and assumptions provided by NTIA.
- Certified cost estimators at each of the services' Cost Centers worked closely with the various program offices to collect the necessary technical and cost data. The cost estimators compiled and reviewed the program data, identified the appropriate program content affected by each system's relocation, developed cost estimates under the given constraints and assumptions, and internally reviewed the estimates consistent with their standard practices before providing them to CAPE to include in the overall estimate.
- CAPE staff reviewed the services' estimates to ensure they adhered to the provided guidelines for accuracy and consistency, and obtained DOD management approval on its practices and findings.

---

[39]The CAPE group is an internal cost-auditing department that provides cost guidance, methods, and tools to DOD employees for estimating costs.

What DOD Found

DOD estimated that it will cost $12.64 billion to relocate most major radio systems operating in the 1755-1850 MHz band. DOD officials said that most systems could be relocated within 10 years, with complete relocation of the satellite systems achieved by 2035. In addition, some systems could relocate from or share the 1755-1850 MHz band within 5 years as a transition step to relocating out of the entire band within the 10-year timeframe. This would require exclusion zones at some locations. Further, for the electronic warfare operations that would continue in the band, enhanced coordination and improved operation procedures would be required.

According to DOD officials, CAPE based this methodology on the cost estimation best practices it customarily employs, revising those practices to suit the study requirements as outlined by NTIA.

We reviewed DOD's preliminary cost estimation methodology and evaluated it against GAO's *Cost Estimating and Assessment Guide* (*Cost Guide*), which also identifies cost estimating best practices, including those used throughout the federal government and industry. The best practices identified in the *Cost Guide* help ensure that cost estimates are comprehensive, well-documented, accurate, and credible. These characteristics of cost estimates help minimize the risk of cost overruns, missed deadlines, and unmet performance targets:

- A *comprehensive* cost estimate ensures that costs are neither omitted nor double counted.
- A *well-documented* estimate is thoroughly documented, including source data and significance, clearly detailed calculations and results, and explanations for choosing a particular method or reference.
- An *accurate* cost estimate is unbiased, not overly conservative or overly optimistic, and based on an assessment of most likely costs.
- A *credible* estimate discusses any limitations of the analysis from uncertainty or biases surrounding data or assumptions.

When applying GAO's identified best practices to DOD's methodology, we took into account that DOD officials developed the preliminary cost estimate for relocation as a less rigorous, "rough order of magnitude" cost estimate,[40] not a budget-quality cost estimate. The nature of a rough-order-of-magnitude estimate means that it is not as robust as a detailed, budget quality life-cycle estimate and its results should not be considered or used with the same level of confidence. Because of this, we performed a high-level analysis of DOD's preliminary cost estimate and methodology, and did not review all supporting data and analysis.

When we reviewed DOD's preliminary cost estimation methodology and evaluated it against the *Cost Guide's* best practices, we found that DOD's methodology substantially met the comprehensive and well-documented

---

[40]The rough-order-of-magnitude estimate is typically developed to support "what-if" analyses and is helpful in examining differences in high-level variation alternatives to see which are most feasble. Because it is developed from limited data and in a short time, it should never be considered a budget-quality cost estimate.

characteristics of reliable cost estimates, and partially met the accurate and credible characteristics, as shown in table 2.

Table 2: Summary Assessment of DOD Spectrum Relocation Cost Estimate Compared to GAO-Identified Best Practices

Characteristic	Best practices supporting the characteristic	GAO's assessment[a] of DOD's practices
Comprehensive		*Substantially met*
	• Estimate includes all costs for the respective programs' entire life cycles, completely defines the programs, and is technically reasonable. • Listing of discrete tasks required to relocate systems (i.e., "work breakdown structure") is product-oriented and at an appropriate level of detail to ensure cost elements are neither omitted nor double-counted. • All cost-influencing ground rules and assumptions are documented.	• Estimates contributing to the report include all costs for the respective programs' entire life cycles including development, production, and maintenance and disposal; define the programs; and are technically reasonable. • Standardized, product-oriented work breakdown structures break out summary costs at an appropriate level of detail to ensure cost elements are neither omitted nor double-counted; however, not all of the programs included a documented work breakdown structure. • Overarching, cost-influencing assumptions that applied to all programs included the schedule for relocating, the bands to be vacated, and inflation rates used; however, not all the individual programs included in the study had evidence of cost-influencing ground rules and assumptions.
Well-documented		*Substantially met*
	• Documentation discusses the technical baseline description; the baseline data are consistent with the estimate. • Documentation captures source data, their reliability, and how they were normalized. • Documentation describes in sufficient detail the calculations performed and the estimating methodology used to derive element's costs. • Documentation describes how the estimate was developed so that a cost analyst unfamiliar with the program could understand what was done and replicate the estimate. • Documentation provides evidence that the cost estimate was reviewed and accepted by management.	• Documentation discusses the technical baseline description; the baseline data are consistent with the estimate. • Documentation captures varying levels of detail on source data consistent with the magnitude of each system's relocation cost (i.e., systems with higher estimated relocation costs are supported with a greater level of detail). • Documentation describes with varying levels of detail the calculations performed and the estimation methodology used; some documentation was not sufficient. • Analysts at each of the service's Cost Centers conducted sufficiency reviews of the cost estimates from the program offices. Documentation for the majority of programs was sufficient such that an analyst unfamiliar with the program could understand and replicate what was done, but some programs did not have detailed data and documentation of the program cost estimates. • Relocation cost estimates were reviewed and accepted by management.

Characteristic	Best practices supporting the characteristic	GAO's assessment[a] of DOD's practices
Accurate	• Estimate results are unbiased, not overly conservative or optimistic, and are based on an assessment of most likely costs. • Estimate has been adjusted properly for inflation. • Estimate contains few, if any, minor errors. • Estimate is based on a historical record of cost estimating and actual experiences from other comparable programs.	*Partially met* • No confidence level was specifically stated in DOD's study to determine if the costs are the most likely costs.[b] • The appropriate inflation rates (OSD 2012 rates) are applied properly. Estimate is properly adjusted and presented in then-year and base-year 2011 dollars. • No calculation errors were apparent. • Estimated costs agree with the historical relocation cost estimate for this band, which DOD compiled in 2001.[c]
Credible	• Estimate's sensitivity analysis identifies a range of possible costs based on varying major assumptions, parameters, and data inputs. • Risk and uncertainty analyses quantify imperfectly understood risks and identify the effects of changing cost-driver assumptions and factors. • Major cost elements were crosschecked to see whether results are similar.	*Partially met* • A sensitivity analysis to identify the range of possible costs at the summary level was not performed by CAPE, although some individual programs completed sensitivity analyses on the range of possible costs. • Risk assessments were completed for only a few of the programs, and not at all by CAPE at the summary level.[d] Risk assessments are needed because the cost estimate is highly dependent on key assumptions, such as the primary band for each system's relocation. However, due to time and resource constraints, no attempt was made to estimate costs to relocate to alternative bands. • Major cost elements were crosschecked with similar results.

Source: GAO Analysis of DOD feasibility study for the 1755-1850 MHz band relocation.

[a]GAO's *Cost Guide* includes five levels of compliance with its best practices. Not Met: Provided no evidence that satisfies any of the characteristic. Minimally Met: Provided evidence that satisfies a small portion of the characteristic. Partially Met: Provided evidence that satisfies about half of the characteristic. Substantially Met: Provided evidence that satisfies a large portion of the characteristic. Fully Met: Provided complete evidence that satisfies the entire characteristic.

[b]Showing that the costs are the most likely costs is required to fully or substantially meet this characteristic.

[c]CAPE compared the overall cost estimate using constant fiscal year 2011 dollars with DOD's 2001 cost estimate for relocating from the same band (Department of Defense, Investigation of the Feasibility of Accommodating the International Mobile Telecommunications (IMT) 2000 Within the 1755-1850 MHz Band (February 9, 2001)), adjusting for changes in the types and quantities of systems, and demonstrated that the two estimates are within 5 percent of each other.

[d]Risk assessments and sensitivity analyses are required on all projects and at the summary level to fully meet this characteristic and on a majority of projects to substantially meet this characteristic. A risk assessment identifies the factors underlying an estimate that might be uncertain and the risks they pose to the estimate. A sensitivity analysis examines how changes to key assumptions and inputs affect the estimate.

Overall, we found that DOD's cost estimate was consistent with the purpose of the feasibility study, which was to inform the decision making process to reallocate 500 MHz of spectrum for commercial wireless broadband use. Additionally, we found that DOD's preliminary cost-estimation methodology substantially met both the comprehensive and well-documented characteristics. As noted in the table above, we observed that DOD's estimate included complete information about systems' life cycles and was generally well-documented. However, these characteristics were not fully met because we found that information on the tasks required to relocate some systems was incomplete, and that documentation for some programs was not sufficient to support a rough-order-of-magnitude estimate. We also determined that DOD's preliminary cost-estimation methodology partially met the accurate and credible characteristics. We found that DOD properly applied appropriate inflation rates and made no apparent calculation errors, and that the estimated costs agree with DOD's prior relocation cost estimate for this band conducted in 2001. However, DOD did not fully or substantially meet the accurate and credible characteristics because it was not clear if the estimate considered the most likely costs and because some sensitivity analyses and risk assessments were only completed at the program level for some programs, and not at all at the summary level.

As the Assumptions Supporting DOD's Cost Estimate for Relocating from the 1755-1850 MHz Band Change, Costs May Also Change

Even though DOD's preliminary cost estimate substantially met some of our best practices, as the assumptions supporting the estimate change over time, costs may also change. According to DOD officials, any change to key assumptions about the bands to which systems would move and the relocation start date could substantially change relocation costs. Because decisions about the spectrum bands to which the various systems would be reassigned and the time frame for relocation have not been made yet, DOD based its current estimate on the most likely assumptions, provided by NTIA, some of which have already been proven inaccurate or are still undetermined. For example:

- **Relocation bands:** Decisions about which comparable or alternate spectrum bands federal agencies, including DOD, should relocate to are still unresolved. According to DOD officials, equipment relocation costs vary significantly depending on the relocation band's proximity to the current band. Moving to bands further away than the assumed relocation bands could increase costs relative to moving to closer bands with similar technical characteristics. In addition, congestion, in both the 1755-1850 MHz band and some of the potential alternate spectrum bands to which federal systems might be moved, complicates relocation planning. According to DOD officials, many of the federal radio systems relocated from the 1710-1755 MHz band were simply re-tuned or compressed into the 1755-1850 MHz band, adding to the complexity of systems and equipment requiring relocation from this band since 2001. Also, DOD officials said that some of the potential spectrum bands to which DOD's systems could be relocated are themselves either already congested or the systems are incompatible unless other actions are also taken. For example, cost estimates for several of DOD's systems assumed that these systems would be relocated into the 2025-2110 MHz band, and operate within this band on a primary basis. However, this band is currently allocated to commercial electronic news gathering systems and other commercial and federal systems, and while the band is not currently congested, it does not support compatible coexistence between DOD systems and commercial electronic news gathering systems. To accommodate military systems within this band, FCC would need to withdraw this spectrum from commercial use to allow NTIA to provide DOD primary status within this band, or FCC would have to otherwise ensure that commercial systems operate on a non-interference basis with military systems. FCC has not initiated a rulemaking procedure to begin such processes.

- **Relocation start date:** DOD's cost estimate assumed relocation would begin in fiscal year 2013, but no auction has been approved, so relocation efforts have not begun. According to DOD officials, a change in the start date creates uncertainty in the cost estimate because new equipment and systems continue to be deployed in and designed for this band, and older systems are retired. This changes the overall profile of systems in the band, a change that can alter the costs of relocation. For example, a major driver of the cost increase between DOD's 2001 and 2011 relocation estimates for the 1755-1850 MHz band was the large increase in the use of the band, including unmanned aerial systems. DOD deployed these systems very little in 2001, but their numbers had increased substantially by 2011. Conversely, equipment near the end of its life cycle when DOD's 2011 relocation cost estimate was completed may be retired or replaced outside of relocation efforts, which could decrease relocation costs.

- **Inflation:** DOD appropriately used 2012 inflation figures in its estimate, assuming that relocation would begin in fiscal year 2013. As more time elapses before the auction occurs, the effect of inflation will increase the relocation costs each year.

According to DOD, the preliminary cost estimate is not as robust as a detailed, budget-quality lifecycle estimate. A budget-quality estimate is based on more fully formed assumptions for specific programs. DOD officials said that for a spectrum relocation effort, a detailed, budget-quality cost estimate would normally be done during the transition-planning phase once a spectrum auction has been approved and would be based on the requirements for the specific auction and relocation decisions.

No Government Revenue Forecasts Exist for a Potential Auction of the 1755-1850 MHz Band, and a Variety of Factors Could Influence Auction Revenues

Federal Agencies Have Not Produced a Revenue Forecast for the 1755-1850 MHz Band

No official government revenue forecast has been prepared for a potential auction of 1755-1850 MHz band licenses, but some estimates might be prepared once there is a greater likelihood of an auction. Officials we spoke with at CBO, FCC, NTIA, and OMB confirmed that none of these agencies has produced a revenue forecast thus far. Officials at these agencies knowledgeable about estimating spectrum-license auction revenue said that because the value of licensed spectrum varies greatly over time and the information on factors that might influence the spectrum auction revenues is not yet available, it is too early to produce meaningful forecasts for a potential auction of the 1755-1850 MHz band. Moreover, CBO only provides written estimates of potential receipts when a congressional committee reports legislation invoking FCC auctions. OMB would estimate receipts and relocation costs as part of the President's Budget; OMB analysts would use relocation cost information from NTIA to complete OMB's estimate of receipts. The potential for large differences between CBO and OMB forecasts exist, as well. For example, in the past, CBO and OMB have produced very different estimates of potential FCC auction receipts at approximately the same time with access to the same data, underscoring how differing assumptions can lead to different results.[41]

Although no official government revenue forecast exists, an economist with the Brattle Group, an economic consulting firm, published a revenue forecast in 2011 for a potential auction of the 1755-1850 MHz band that forecasted revenues of $19.4 billion for the band.[42] We did not evaluate the accuracy of this revenue estimate. Like all forecasts, the Brattle Group study was based on certain assumptions. For example, it assumed that the band would generally be cleared of federal users. It also assumed the AWS-1 average nationwide price of $1.03 per "MHz-pop" as a baseline price for spectrum allocated to wireless broadband services.[43] In addition, the study adjusts the price of spectrum based on the following considerations:

- *Increase in the quantity of spectrum using elasticity of demand.* As the supply of spectrum for commercial wireless broadband services increases, the price and value of spectrum is expected to fall. The elasticity of demand is used to make adjustments for the increased supply of spectrum.[44]
- *Differences in capacity and quality of spectrum using value weights.* The study assigns value weights to different bands of spectrum compared to the quality of the AWS-1 band. A lower value weight is given for spectrum that is not symmetrically paired, for example,

---

[41]In 1997, OMB included in the President's budget request for fiscal year 1998 an estimate that spectrum auctions, including proposals that were subject to policy changes, would raise $36.1 billion between 1998 and 2002. CBO estimated that the same basic policy proposals would raise roughly two-thirds as much, $24.3 billion, over the same period. CBO, *Where Do We Go From Here? The FCC Auctions And The Future of Radio Spectrum Management* (Washington, D.C.: April 1997). Actual winning bids for spectrum auctions between fiscal years 1998 and 2002 were $19.5 billion.

[42]Coleman Bazelon, The Brattle Group, Inc., *Expected Receipts From Proposed Spectrum Auctions* (Washington, D.C.: July 28, 2011).

[43]The unit price of licensed spectrum is typically expressed in terms of dollars per "MHz-pop," where MHz-pop is the product of total MHz of a band and population covered by the region of a license. The $1.03 price represents the current price for AWS-1 spectrum based on the original AWS-1 auction price adjusted for inflation using the SpecEx Spectrum Index.

[44]According to the study, wireless broadband spectrum is generally thought to have a price elasticity of around -1.2, which implies that a 1 percent increase in the base supply of spectrum should result in a 1.2 percent decrease in its price.

because traditional, two-way communications, such as mobile phone services, are typically provided over paired bands of spectrum.[45] Similarly, a greater value weight is given to bands of spectrum with no restrictions on use, or encumbrances. Fewer restrictions would increase the capacity or the types of services for a given spectrum band.

The study also assumes that the 1755-1780 MHz portion of the band is paired with the 2155-2180 MHz band, which various industry stakeholders currently support. For spectrum services that require two-way communications, pairing bands allows them to be used more efficiently by diminishing interference from incompatible adjacent operations.[46] In addition, the study assumed the 95 MHz of spectrum between 1755 and 1850 MHz would be auctioned as part of a total of 470 MHz of spectrum included in six auctions sequenced 18 months apart and spread over 9 years with total net receipts of $64 billion. Thus, the forecast also took into account when the spectrum would be reallocated for commercial services.

A Variety of Factors Ultimately Influence Auction Revenue

Like all goods, the price of licensed spectrum, and ultimately the auction revenue, is determined by supply and demand. This fundamental economic concept helps to explain how the price of licensed spectrum could change depending on how much spectrum is available now and in the future, and how much licensed spectrum is demanded by the wireless industry for broadband applications. Government agencies can influence the supply of spectrum available for licensing and the characteristics of those licenses, whereas expectations about profitability determine demand for spectrum in the marketplace.[47]

---

[45]With a paired spectrum band, a portion of the frequencies (usually half) are used to transmit from a base station to a mobile device, and the remainder of the band is used for mobile to base station transmissions. Newer technologies now allow for the use of unpaired spectrum for two-way communications.

[46]See Bazelon, *The Economic Basis of Spectrum Value.*

[47]The value of a spectrum license and hence the future price of licensed spectrum at a given auction depends on many factors, ranging from the propagation characteristics of the particular spectrum to the general investment climate and the existence of applicable technology infrastructure. For the purposes of this discussion, we focus only on those supply and demand factors directly influenced by government decisions or wireless companies.

Supply. FCC and NTIA, with direction from Congress and the President, jointly influence the amount of spectrum allocated for federal and nonfederal users, including the amount to be shared by federal and nonfederal users. In 2010, the President directed NTIA to work with FCC to make 500 MHz of spectrum available for use by commercial broadband services within 10 years. This represents a significant increase in the supply of spectrum available for licensing in the marketplace. As with all economic goods, with all other things being equal, the price and value of spectrum licenses are expected to fall as additional supply is introduced. However, at this time, the answers to key questions about the reallocation of the 1755-1850 MHz band are unknown. Expectations about exactly how much spectrum is available for licensing now and how much will be available in the future would influence how much wireless companies would be willing to pay for spectrum licensed today.

Demand. The expected, potential profitability of a spectrum license influences the level of demand for it. As with all assets, companies base their capital investment decisions on the expected net return, or profit, over time of their use. The same holds true for spectrum. Currently, the demand for licensed spectrum is increasing, and a primary driver of this increased demand is the significant growth in the use of commercial-wireless broadband services, including third and fourth generation technologies that are increasingly used for smart phones and tablet computers. Below are some of the factors that would influence the demand for licensed spectrum:

- *Clearing versus Sharing*: Spectrum is more valuable, and companies will pay more to license it, if it is entirely cleared of incumbent federal users, giving them sole use of licensed spectrum; spectrum licenses are less valuable if access must be shared. Sharing could potentially have a big impact on the price of spectrum licenses, especially if a sharing agreement does not guarantee service when the licensee would need it most. For example, knowing in advance that service would be unavailable once a month at 3 a.m. may not significantly influence price, but if the times when the service will be unavailable are unknown, the effect on price could be significant. In 2012, the President's Council of Advisors on Science and Technology advocated that sharing between federal and commercial users become the new norm for spectrum management, especially given the high cost and lengthy time it takes to relocate federal users and the disruptions to agencies' missions.

- *Certainty and Timing*: Another factor that affects the value of licensed spectrum is the certainty about when it becomes available. Seven years after the auction of the 1710-1755 MHz band, federal agencies are still relocating systems. According to an economist with whom we met, one lesson from the 1710-1755 MHz relocation effort is that uncertainty about the time frame for availability reduces the value of the spectrum. Any increase in the probability that the spectrum would not be cleared on time would have a negative impact on the price companies are willing to pay to use it. As such, the estimated 10-year timeframe to clear federal users from the entire 1755-1850 MHz band, and potential uncertainty around that time frame, could negatively influence demand for the spectrum. The 2012 amendments to the CSEA include changes designed to reduce this uncertainty by requiring federal agencies that will be relocating (or sharing spectrum) to submit transition plans with timelines for specific geographic locations, with interagency review of those plans aimed at ensuring timely relocation (or sharing) arrangements.

- *Available Wireless Services*: Innovation in the wireless broadband market is expected to continue to drive demand for wireless services. For example, demand continues to increase for smart phones and tablets as new services are introduced in the marketplace. These devices can connect to the Internet through regular cellular service using commercial spectrum, or they can use publicly available (unlicensed) spectrum via Wi-Fi networks to access the Internet.[48] The value of the spectrum, therefore, is determined by continued strong development of and demand for wireless services and devices, and the profits that can be realized from them.

Agency Comments

We provided a draft of this report to the Department of Commerce (Commerce), DOD, FCC, and OMB for review and comment. FCC agreed with the report's findings, and Commerce, DOD, and FCC provided technical comments that we incorporated as appropriate. FCC's written comments appear in appendix II. OMB did not provide comments.

We are sending copies of this report to the Secretary of Commerce, the Secretary of Defense, the Chairman of the Federal Communications Commission, the Director of the Office of Management and Budget, and

---

[48] Wi-Fi networks can permit multiple computing devices in each discrete location to share a single wired connection to the Internet, thus efficiently sharing spectrum.

the appropriate congressional committees. In addition, the report will be available at no charge on GAO's website at http://www.gao.gov.

If you or members of your staff have any questions about this report, please contact me at (202) 512-2834 or goldsteinm@gao.gov. Contact points for our Offices of Congressional Relations and Public Affairs may be found on the last page of this report. Major contributors to this report are listed in appendix III.

Mark L. Goldstein
Director
Physical Infrastructure Issues

Appendix I: Objectives, Scope, and Methodology

The objectives of this report were to examine (1) the differences, if any, between estimated and actual federal relocation costs and auction revenues from the 1710-1755 MHz band; (2) the extent to which the Department of Defense (DOD) followed best practices to prepare its preliminary cost estimate for vacating the 1755-1850 MHz band, and any limitations of its analysis; and (3) what government or industry revenue forecasts for the 1755-1850 MHz band auction exist, if any, and what factors, if any, could influence actual auction revenue.

To examine the differences, if any, between estimated and actual federal relocation costs and auction revenues from the 1710-1755 MHz band, we reviewed spectrum auction data published by the Federal Communications Commission (FCC) and federal relocation cost data from the National Telecommunication and Information Administration's (NTIA) annual 1710-1755 MHz band relocation progress reports, published yearly since 2008.[1] We narrowed our review of past spectrum auctions to the 1710-1755 MHz relocation after reviewing FCC auction data and NTIA reports describing other spectrum relocations and auctions involving federal agencies, and interviews with knowledgeable FCC, NTIA, Office of Management and Budget (OMB), and Congressional Budget Office (CBO) officials. The Advanced Wireless Services-1 (AWS-1) auction involving the 1710-1755 MHz band is the only spectrum auction involving federal agencies with significant, known relocation costs.[2] In addition, it is the only relocation involving DOD radio communication systems. To assess the reliability of FCC auction and NTIA relocation cost data, we reviewed documentation related to the data; compared it to other sources, including other government reports; and discussed the data with FCC and NTIA officials. We did not evaluate the accuracy of individual agencies' relocation cost data, as this was outside the scope of our review. Based on this review, we determined that the FCC and NTIA data were sufficiently reliable for the purposes of our report.

To determine the extent to which DOD followed best practices to prepare its preliminary cost estimate for vacating the 1755-1850 MHz band, we

---

[1]See, for example, NTIA, *Relocation of Federal Radio Systems from the 1710-1755 MHz Spectrum Band; Sixth Annual Progress Report* (Washington, D.C.: March 2013).

[2]There have been other auctions of federally allocated spectrum involving the relocation of federal government agencies, as noted previously.

Appendix I: Objectives, Scope, and Methodology

assessed DOD's preliminary cost estimate against the best practices in GAO's *Cost Estimating and Assessment Guide* (*Cost Guide*), which has been used to evaluate cost estimates across the government.[3] These best practices help ensure cost estimates compiled at different stages in the cost estimating process are comprehensive, well-documented accurate, and credible. To develop our assessment, we interviewed DOD officials, including in the agency's Cost Assessment and Program Evaluation (CAPE) group that led the cost estimation effort, regarding their data collection and cost estimation methodologies and the findings reported in DOD's feasibility study. We also reviewed electronic source documentation supporting the estimate with a CAPE official. After completing this review, a GAO cost analyst developed an assessment using our 5-point scale (not met, minimally met, partially met, substantially met, and met) and a second analyst verified the assessment. DOD's preliminary cost estimate was a rough-order-of-magnitude estimate; consequently, it did not contain all the information expected of a complete, budget-quality cost estimate. Therefore, we performed a high-level analysis to determine whether DOD's reported estimated costs considered all the potential factors that could influence those relocation costs. To identify any limitations affecting DOD's estimate, we interviewed DOD officials responsible for developing the department's preliminary cost estimate. We also interviewed NTIA and OMB officials knowledgeable about the intended purpose of the estimate to discuss how the estimate should be used and any factors that would affect the reliability of the estimate.

To identify any government or industry revenue forecasts from a future auction of licenses in the 1755-1850 MHz band, we reviewed government, industry, and public policy reports, and interviewed officials from CBO, FCC, NTIA, and OMB. We interviewed an economist at the Brattle Group, an economic-consulting firm, who published a paper describing expected receipts from a future auction of 1755-1850 MHz band licenses.[4] To determine any factors that would affect the price of spectrum licenses, we analyzed academic, government, and public policy literature on spectrum valuation. We narrowed our search to those reports or papers specifically mentioning (1) spectrum auctions involving the relocation of federal agencies, (2) spectrum valuation and revenues from

---

[3]GAO-09-3SP.

[4]Bazelon, *Expected Receipts From Proposed Spectrum Auctions* (July 2011).

Appendix I: Objectives, Scope, and Methodology

the sale of spectrum licenses, and (3) relocation costs. We discussed factors affecting spectrum auction revenue with CBO and OMB officials, industry and policy experts, and obtained input from CTIA—The Wireless Association, the association representing the wireless industry.

We conducted this performance audit from September 2012 to May 2013 in accordance with generally accepted government auditing standards. Those standards require that we plan and perform the audit to obtain sufficient appropriate evidence and provide a reasonable basis for our findings and conclusions based on our audit objectives. We believe that the evidence obtained provides a reasonable basis for our findings and conclusions based on our audit objectives.

Appendix II: Comments from the Federal Communications Commission

Federal Communications Commission
Washington, D.C. 20554

May 15, 2013

Mark Goldstein
Director
Physical Infrastructure Issues
Government Accountability Office
441 G Street, NW
Washington, DC 20548

Dear Mr. Goldstein:

Thank you for the opportunity to review GAO's draft report Spectrum Management: Federal Relocation Costs and Auction Revenues (GAO-13-472).

We agree with the draft report's findings and conclusions regarding the FCC's role in this issue.

Sincerely,

Ruth Milkman
Chief, Wireless Telecommunications Bureau

Appendix III: GAO Contact and Staff Acknowledgments

GAO Contact	Mark L. Goldstein, (202) 512-2834 or goldsteinm@gao.gov
Staff Acknowledgments	In addition to the contact named above, Michael Clements, Assistant Director; Stephen Brown; Jonathan Carver; Leia Dickerson; Jennifer Echard; Emile Ettedgui; Colin Fallon; Bert Japikse; Elke Kolodinski; Joshua Ormond; Jay Tallon; and Elizabeth Wood made key contributions to this report.

(543315)

GAO's Mission	The Government Accountability Office, the audit, evaluation, and investigative arm of Congress, exists to support Congress in meeting its constitutional responsibilities and to help improve the performance and accountability of the federal government for the American people. GAO examines the use of public funds; evaluates federal programs and policies; and provides analyses, recommendations, and other assistance to help Congress make informed oversight, policy, and funding decisions. GAO's commitment to good government is reflected in its core values of accountability, integrity, and reliability.
Obtaining Copies of GAO Reports and Testimony	The fastest and easiest way to obtain copies of GAO documents at no cost is through GAO's website (http://www.gao.gov). Each weekday afternoon, GAO posts on its website newly released reports, testimony, and correspondence. To have GAO e-mail you a list of newly posted products, go to http://www.gao.gov and select "E-mail Updates."
Order by Phone	The price of each GAO publication reflects GAO's actual cost of production and distribution and depends on the number of pages in the publication and whether the publication is printed in color or black and white. Pricing and ordering information is posted on GAO's website, http://www.gao.gov/ordering.htm. Place orders by calling (202) 512-6000, toll free (866) 801-7077, or TDD (202) 512-2537. Orders may be paid for using American Express, Discover Card, MasterCard, Visa, check, or money order. Call for additional information.
Connect with GAO	Connect with GAO on Facebook, Flickr, Twitter, and YouTube. Subscribe to our RSS Feeds or E-mail Updates. Listen to our Podcasts. Visit GAO on the web at www.gao.gov.
To Report Fraud, Waste, and Abuse in Federal Programs	Contact: Website: http://www.gao.gov/fraudnet/fraudnet.htm E-mail: fraudnet@gao.gov Automated answering system: (800) 424-5454 or (202) 512-7470
Congressional Relations	Katherine Siggerud, Managing Director, siggerudk@gao.gov, (202) 512-4400, U.S. Government Accountability Office, 441 G Street NW, Room 7125, Washington, DC 20548
Public Affairs	Chuck Young, Managing Director, youngc1@gao.gov, (202) 512-4800 U.S. Government Accountability Office, 441 G Street NW, Room 7149 Washington, DC 20548

Please Print on Recycled Paper.